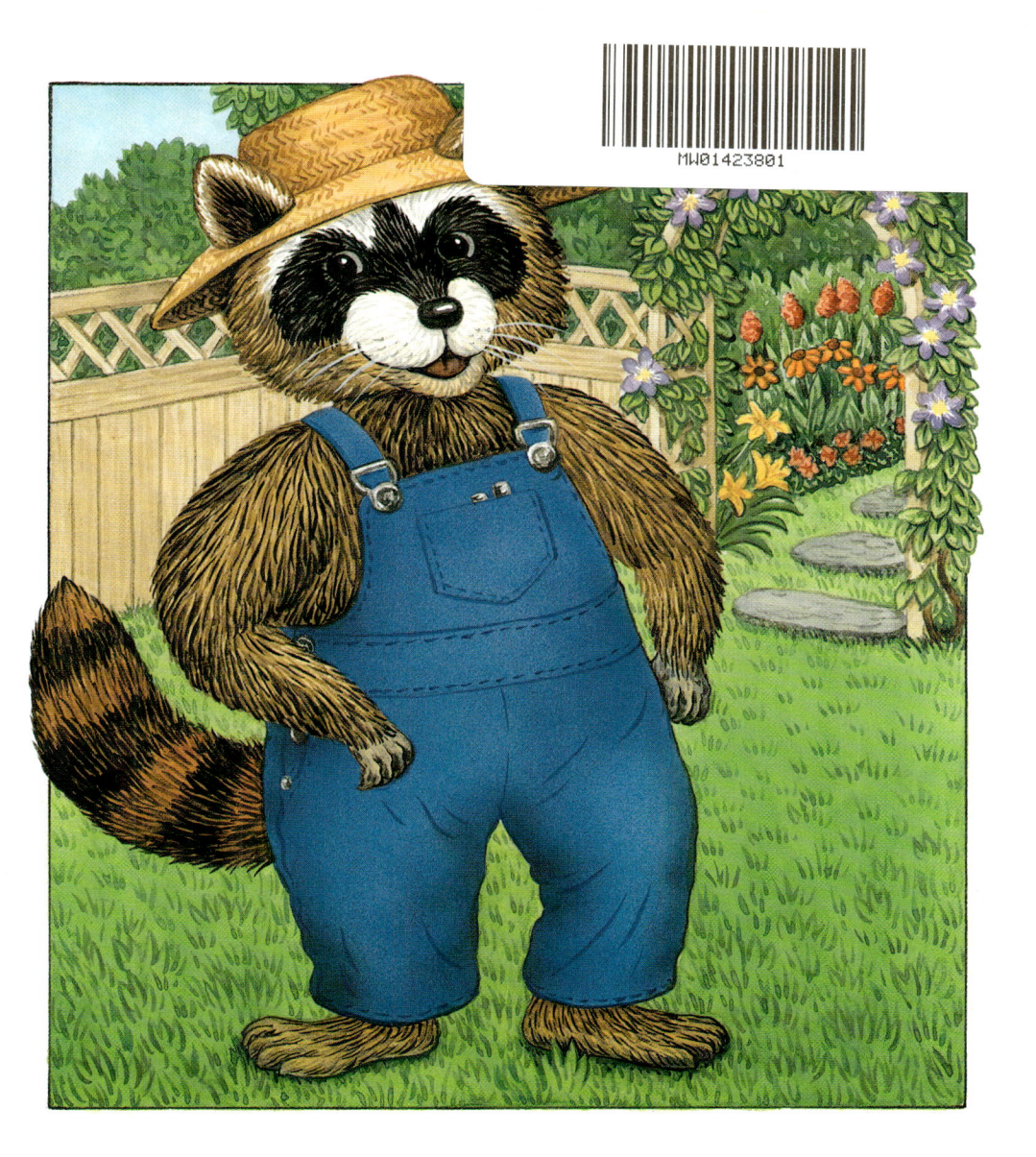

I am Reed the Raccoon.

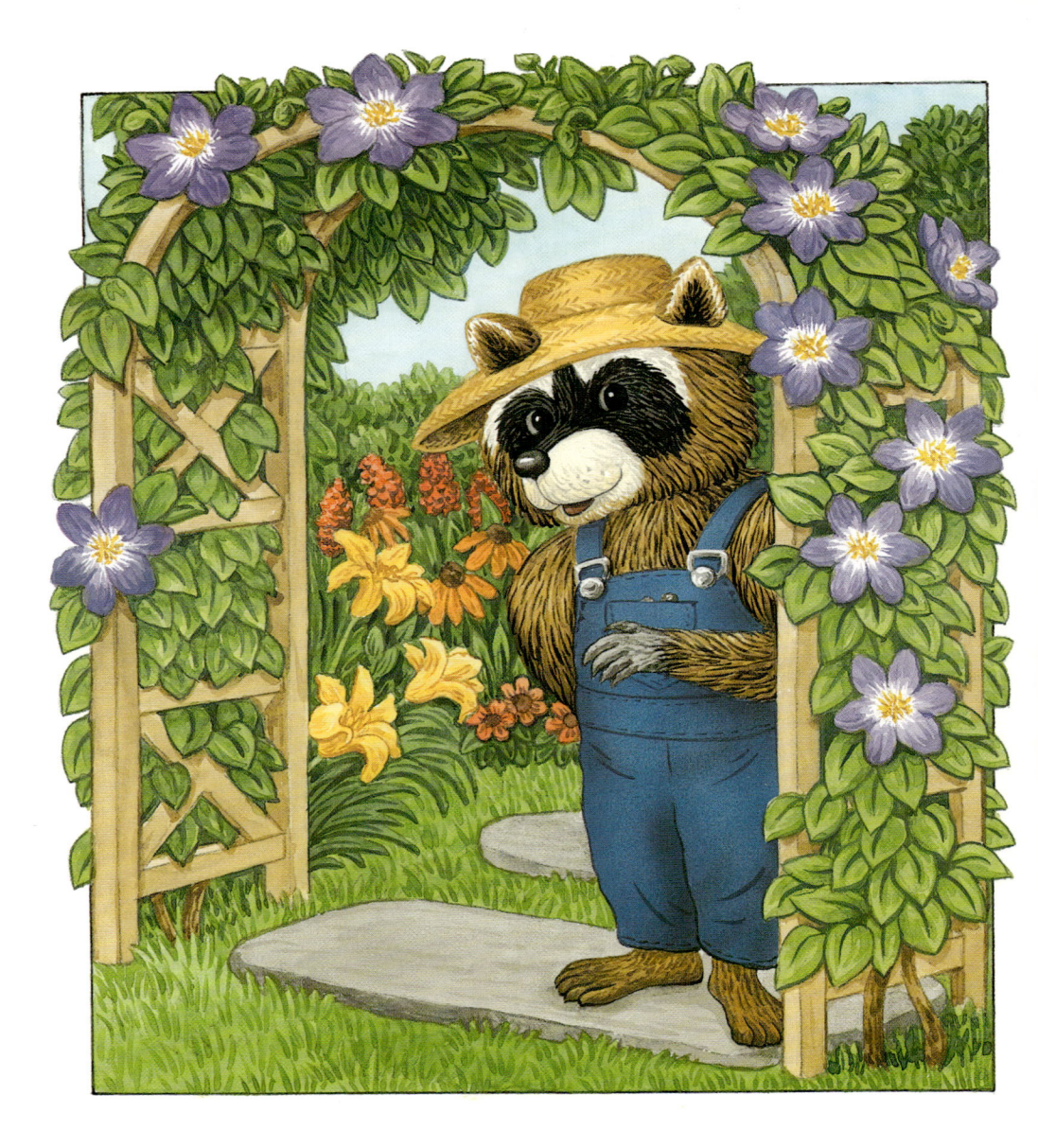

Come and see the garden.

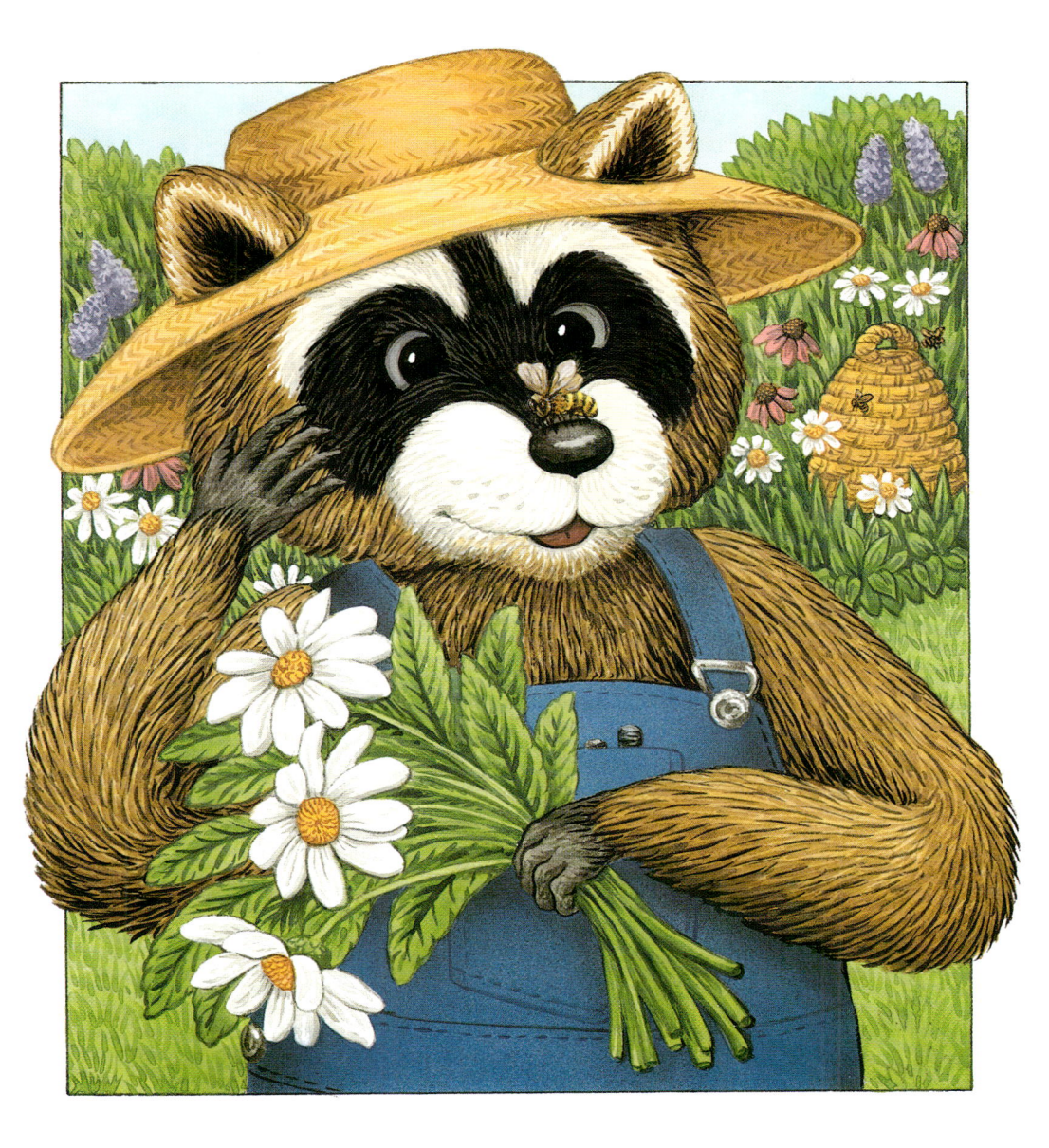

Is that a bee I see?

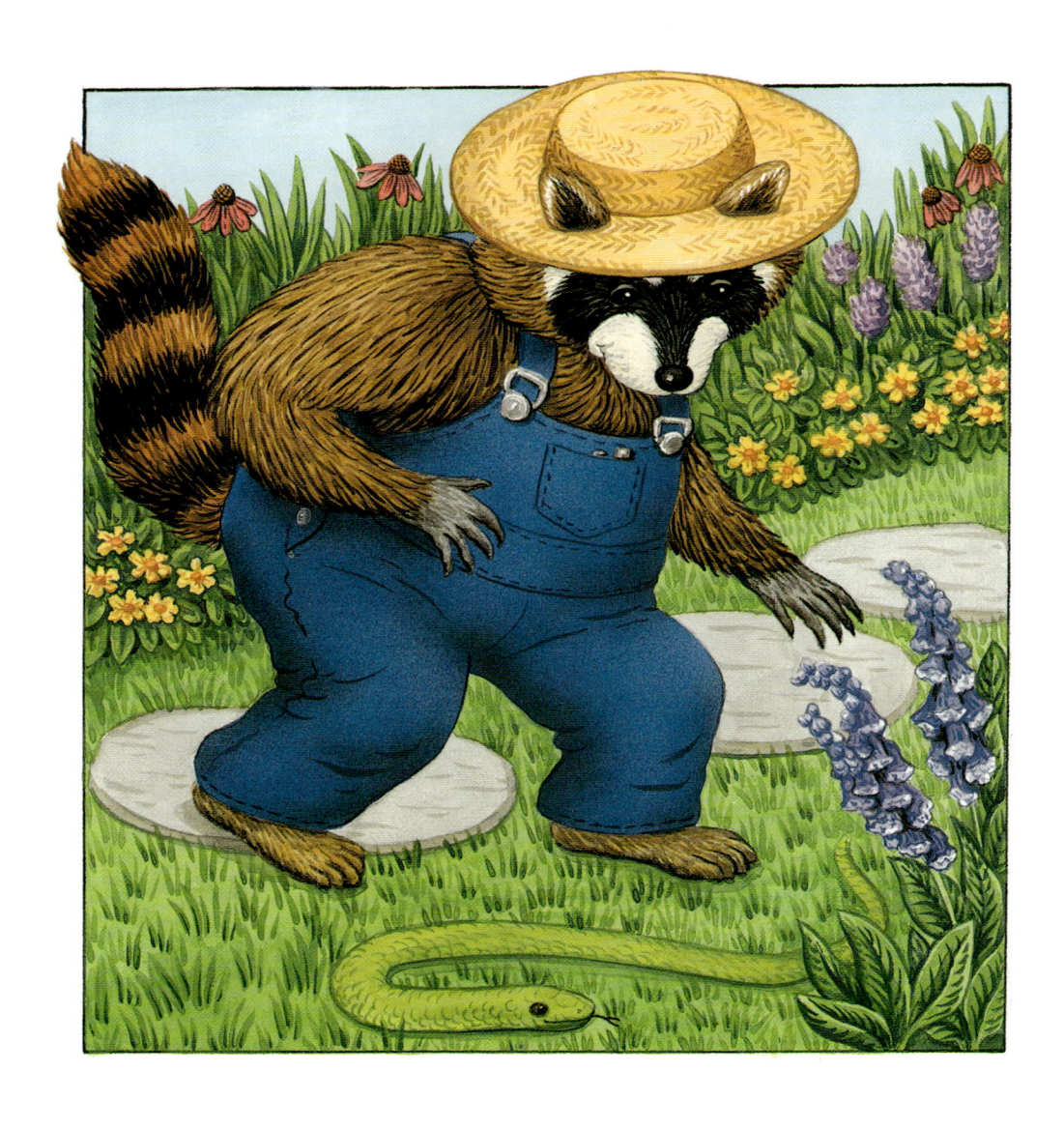

See the garden snake. It is green.

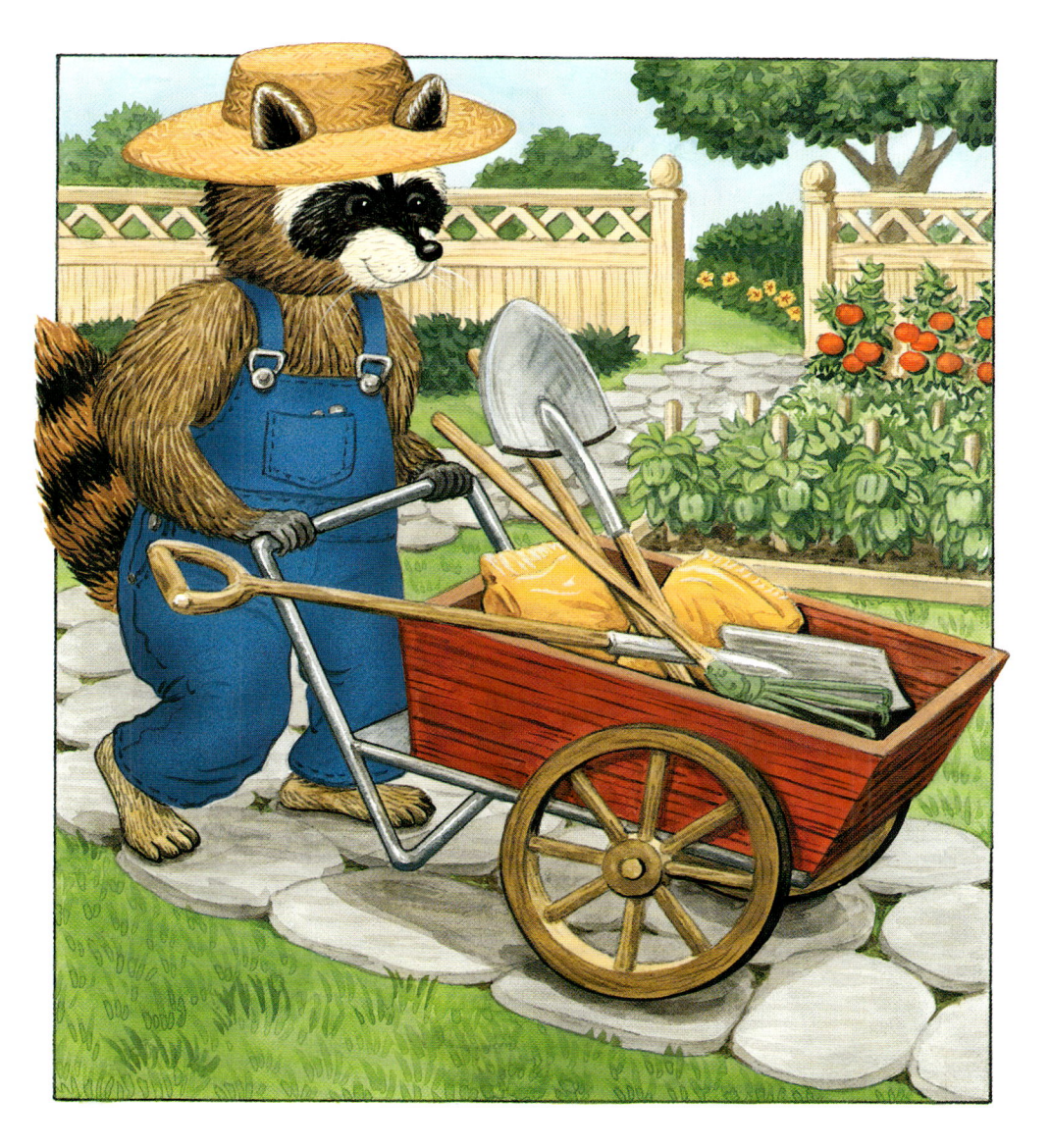

I keep tools I need.

I plant three trees.

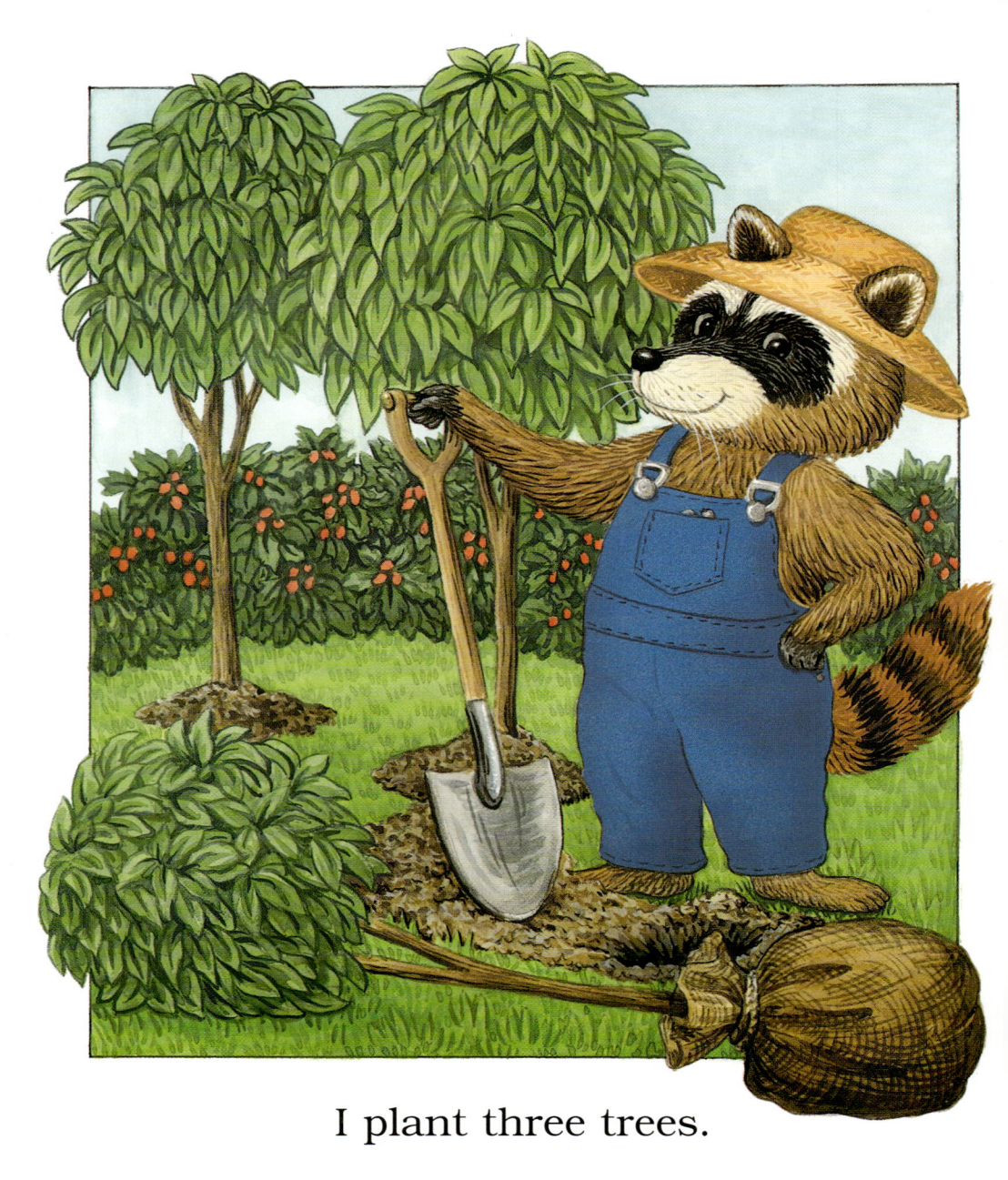

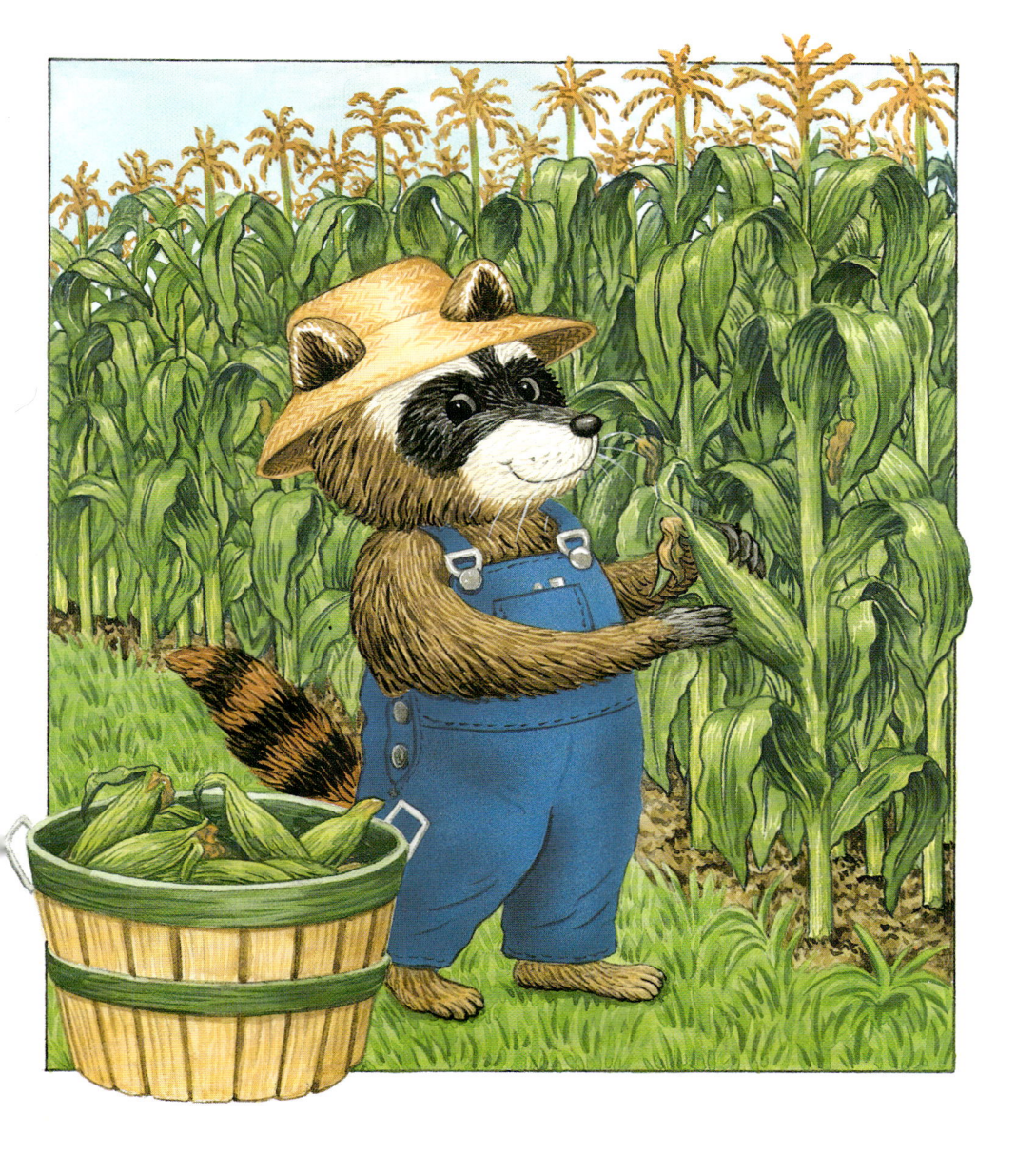

I pick a basket of sweet corn.

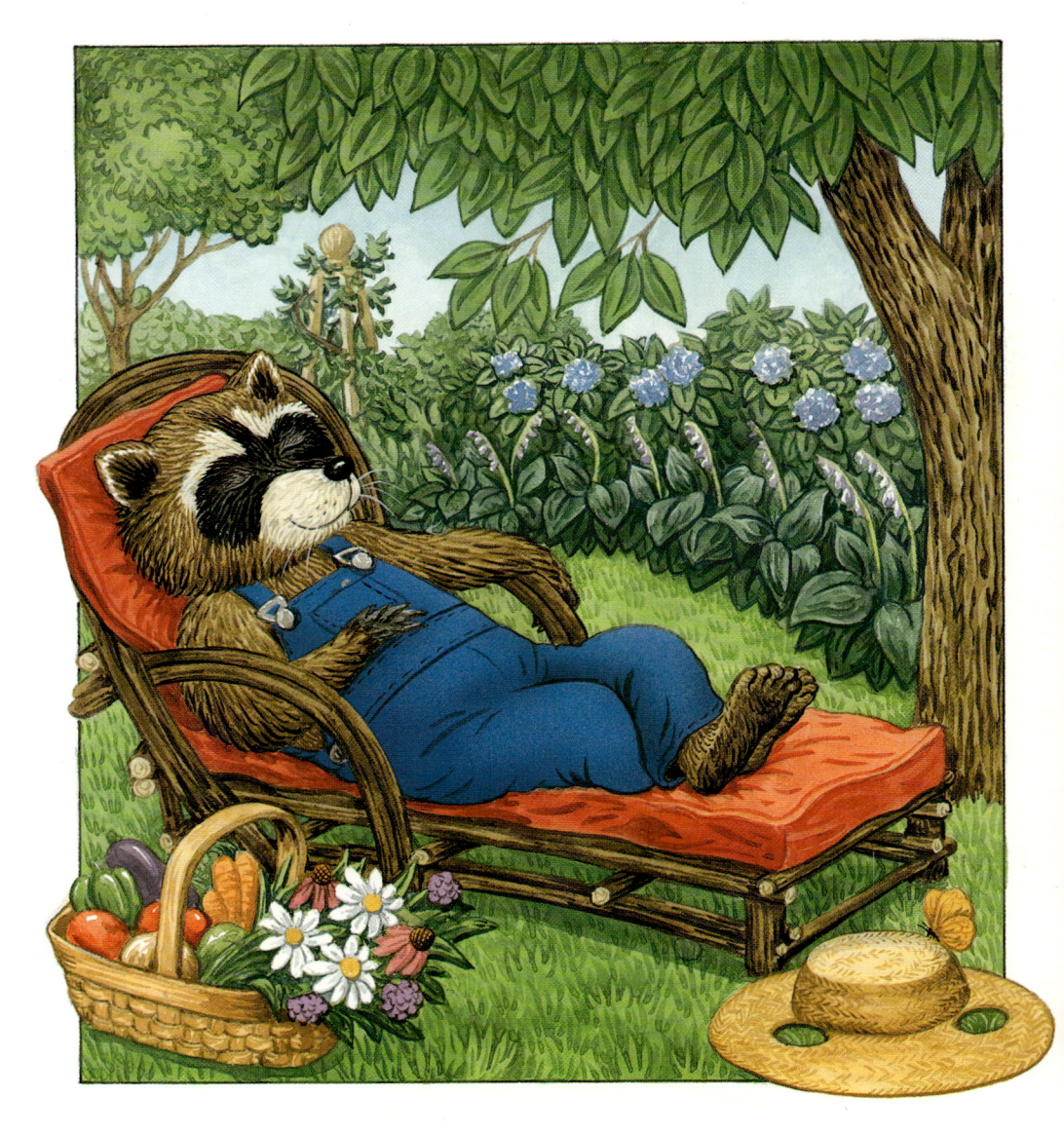

I need some sleep!